Altri miei libri di possibile interesse:

Serie: panoramica tecnologie. https://bit.ly/3KiIVh4

Serie: capire la Scienza per tutti. https://bit.ly/3UfgbdD

Intelligenza Artificiale (Basi).(https://amzn.eu/d/52yrt2N

INTELLIGENZA ARTIFICIALE AL LAVORO

Come utilizzare l'AGI-AI, la generazione di immagini, porre domande, ottenere descrizioni, diritti d'autore, 134 esempi.

INTELLIGENZA ARTIFICIALE AL LAVORO

Come utilizzare l'AGI-AI, la generazione di immagini, porre domande, ottenere descrizioni, diritti d'autore, 134 esempi.

Ettore Accenti

Blog: http://ettoreaccenti.blogspot.ch/

DISTIWORLD PUBLISHING

INTELLIGENZA ARTIFICIALE AL LAVORO

Come utilizzare l'AGI-AI, la generazione di immagini, porre domande, ottenere descrizioni, diritti d'autore, 134 esempi.

ISBN-13: 9798869692733
Distiworld Publishing

Dedica

A mia moglie Eva che ha corretto il testo e scelto le immagini

AUTHOR AT CERN'S LHC (GENEVA) WHERE HIGGS BOSON WAS DISCOVERED

Sono un noto autore di testi tecnici e scientifici rivolti a studenti e appassionati di tali argomenti. Utilizzo un mio metodo per rendere comprensibili anche argomenti difficili senza rinunciare al necessario rigore, metodo che amo definire "divulgazione quasi-popolare". Utilizzo un po' di matematica dove indispensabile per completare concetti che lo richiedono.

Ho collaborato con riviste tecniche per decenni, pubblicando una vasta gamma di argomenti oggi diventati di grande attualità e che rappresentano conoscenze fondamentali per chiunque voglia rimanere al passo con i progressi della nostra civiltà.

Con una laurea in Ingegneria Elettrotecnica e decenni di esperienza nel mondo della tecnologia ed avendo lavorato sia in Europa che nella Silicon Valley ho acquisito una panoramica unica su ciò che l'umanità sta sviluppando. Questa mia lunga esperienza in campo mi ha fornito ispirazione per le mie numerose opere.

SUMMARY

Sommario

Premessa

Di tecnologia scrivo da quando era alla terza media ed è veramente stupefacente constatare come il comportamento umano sia sempre lo stesso!

Ho vissuto la nascita dell'informatica, ammirato quei maestosi e terrificanti computer della prima era, l'arrivo dei semiconduttori, la realizzazione del primo microprocessor, l'era del personal computer, lo sviluppo dell'intelligenza artificiale e recentemente il computer quantistico super veloce. Per completare il mio scenario è arrivata ora l'intelligenza artificiale per tutti che sta provocando una marea di perplessità senza che ci sia praticamente nulla di nuovo.

Anticipo subito che ho cominciato ad utilizzarla appena resasi disponibile al pubblico attraverso sistemi come ChatGPT ed altri con enormi vantaggi per la mia attività di maturo scrittore di articoli e libri tecnologici.

E sapete quale è stato il primo vantaggio che ho riscontrato? Un risparmio di tempo! Nel redigere i miei testi spesso ricorrevo a Google per trovare dati: per esempio se ad un certo punto citavo Albert Einstein e volevo indicarne data di nascita e di morte li cercavo su Google search il quale mi tirava fuori i soliti mille link, tra i quali wikipedia, che aprivo, leggevo alcune righe e ne riportavo i dati; totale, operazione da 2 a 4 minuti. Se quello che cercavo del bravo Einstein fosse stato più complesso ci avrei perso 10-20 minuti.

Con ChatGPT digito "dati anagrafici Albert Einstein", ed ecco l'immediata risposta "Albert Einstein (1879-1955) è stato un fisico teorico tedesco, meglio conosciuto per la sua teoria della relatività, specialmente l'equazione E=mc^2, che ha avuto un profondo impatto sulla fisica moderna. È nato il 14 marzo 1879 a Ulm, nel Regno del Württemberg nell'Impero tedesco, e morì il 18 aprile 1955 a Princeton, nel New Jersey, USA." Tempo un secondo. Già solo questo mi convinse della sua utilità.

Chiaro che ChatGPT può fare ben di più, ma non ha assolutamente nulla di nuovo e nulla di meraviglioso dal punto di vista tecnologico.

Tutto è dovuto all'aumentata velocità di calcolo dei computer e all'aumentata capacità dell'informatica di creare algoritmi sempre più complessi, in sintesi lo dobbiamo alla vecchia "legge di Moore" che Gordon Moore individuò nel 1965: "tutto raddoppia ogni due anni!".

Tutte le nuove applicazioni di IA riescono a pescare in un oceano di dati quelli che corrispondono alla mia domanda in nanosecondi.

Il percorso non è diverso da quanto succede con gli iPhone che milioni di persone possiedono e che si sono evoluti con quella legge dall'iPhone 4.0 all'iPhone 15 di oggi e nessuno se ne meraviglia.

Aggiungo che tutto è ed era prevedibile, come prevedibile è a che punto sarà questa tecnologia nell'anno 2030 o nell'anno 2050 anzi, posso già affermare con un semplice calcolo che nell'anno 2050 il numero di transistor contenibili in un iPhone 35 saranno esattamente quanto le sinapsi del nostro cervello: un milione di miliardi!

E qui devo sottolineare la strana sindrome di noi umani dovuta a quell'innata paura per le cose nuove e che non comprendiamo e per le quali siamo portati a diffidarne da una parte od a idolatrarle come magiche dall'altra.

Scopo di questo libro è proprio quello di sfatare tante dicerie e paure e di informare, soprattutto i giovani, che anche questa conquista della tecnologia è un passo avanti dell'umanità e che va capita, studiata ed applicata nel miglior del modo possibile, come da tempo facciamo con i computer.

Questi nuovi strumenti, che da una parte ci agevolano per certe loro utilità e dall'altra ci possono far paura, posso solo suggerire che è un utile e comodo strumento collaborativo il cui uso non va demonizzato, ma che va spiegato soprattutto ai giovani.

Come sfruttare la IA

Le applicazioni possibili sono immense, si pensi anche solo alla guida autonoma delle auto, agli algoritmi di apprendimento automatico ed alla avveniristica idea di collegare anche fisicamente le funzioni del nostro cervello ad elaboratori esterni. Stiamo assistendo ad una rivoluzione paragonabile a quella dell'energia elettrica di poche generazioni fa e che ha contribuito in modo sostanziale a passare dal miliardo di abitanti agli otto miliardi di oggi, con vita media raddoppiata.

Come chiunque di noi oggi maneggia oggetti elettrici con semplicità e senza paura, così l'intelligenza artificiale sta diffondendosi permettendocene un uso generalizzato in molte nostre attività e lo scopo di questo libro è proprio quello di chiarire con parole chiare e semplici esempi come chiunque possa trarne vantaggio.

Per chi fosse interessato alla storia dell'intelligenza artificiale ed alle sue costituenti tecniche può leggere un mio precedente libro al riguardo(https://amzn.eu/d/52yrt2N).

Di seguito valuteremo due attività che sono alla portata di chiunque e cioè come creare testi ed immagini e lo faremo con esempi senza tanti fronzoli di spiegazioni teoriche: immaginiamo in pratica di essere ritornati alla prima elementare a quando ci veniva insegnato come tenere la semplice penna in mano e con questa creare simboli e figure su un foglio bianco. In fondo, quello che c'è da imparare è tutto qui!

Detto questo e prima di iniziare il nostro excursus sul semplice utilizzo della IA, sappiate che essenzialmente l'intelligenza artificiale in generale si divide nelle seguenti quattro categorie.

Le fondamentali forme dell'IA

Da tempo l'intelligenza artificiale si è evoluta da semplici compiti di classificazione e riconoscimento di modelli a sistemi in grado di utilizzare dati storici per fare previsioni imparando dai dati.

l'IA che guida auto e vince nel gioco del Go, sono definiti IA debole. Questi tipi di IA mancano di intelligenza generale come la si definisce oggi e che si dividono come segue.

IA reattiva (Reactive AI). Usa algoritmi che non usano memoria: dato un input specifico, l'output è sempre lo stesso. I modelli di apprendimento automatico che utilizzano questo tipo di IA sono efficaci per compiti di classificazione e riconoscimento di modelli. Possono considerare enormi blocchi di dati e produrre un output apparentemente intelligente, ma sono incapaci di analizzare scenari che includono informazioni imperfette o richiedono una comprensione storica.

Macchine a memoria limitata (Limited Memory Machines). Si basano sulla nostra comprensione di come funziona il cervello umano e sono progettati per imitare il modo in cui i nostri neuroni si comportano. Gestisce compiti di classificazione complessi e utilizza dati storici per fare previsioni. Esegue compiti complessi come la guida autonoma. Per molte attività superano l'uomo, ma si considera con intelligenza modesta perché richiedono enormi quantità di dati di addestramento per imparare compiti che gli umani imparano con solo pochi esempi.

Teoria della mente (Theory of mind). Conosciuta anche come AGI (Artificial General Intelligence) adottata da OpenAI e che ci riguarda in questo libro. Questo è un tipo di IA che punta a ragionare come gli umani e quindi in grado di fornire risultati personalizzati in base alle esigenze espresse dalla richiesta. Può imparare con pochi esempi rispetto alle macchine a memoria limitata; può contestualizzare e generalizzare le informazioni ed estrapolare la conoscenza per un ampio insieme di problemi. Le capacità comunque sono limitate alle informazioni con cui viene addestrata e non può "pensare" in modo autonomo e tantomeno avere coscienza umana.

IA autoconsapevole (Self-Aware AI o ASI). Conosciuta anche come super intelligenza artificiale. È consapevole di sé e dello stato mentale di altre entità. È definita come una macchina con intelligenza pari all'intelligenza generale umana e in linea di principio capace di superare di gran lunga la capacità cognitiva umana ed in grado di creare versioni sempre più intelligenti di sé stessa. Allo stato attuale il limite consiste nel fatto che siamo ancora lontani dall'effettuare il "Reverse Engineering" del cervello umano.

Mettiamo l'IA al nostro servizio

Tra le nostre molte attività ci sono le due di cui abbiamo accennato e che ci seguono da quando, piccoli discepoli della scuola, abbiamo appreso come scrivere e disegnare con una penna. Allo stesso modo qui apprenderete come scrivere e disegnare sfruttando la IA del modello AGI come fosse la vostra nuova penna.

Dobbiamo partire dal fatto che prima di scrivere o disegnare qualcosa dobbiamo formarci l'idea di quanto vogliamo fare nella nostra mente e poi tradurlo in movimenti della nostra mano per trasferire l'idea sulla carta.

Nel nostro caso la mano opera su una tastiera e l'idea deve essere tradotta in una serie di comandi che la tastiera comunica al programma che elabora la nostra richiesta per trasformarla in un testo od una immagine.

In altre parole, con la moderna IA l'algoritmo non solo trasferisce le nostre lettere esattamente come avviene con Word di Microsoft ma riesce ad interpretarne il significato ed autonomamente creare la nostra idea, a suo modo e in modo autonomo.

Ad esempio, posso digitare "crea una breve poesia che parla di fiori per Maria" e l'IA crea una piccola poesia che possiamo utilizzare.

Se digito "crea un'immagine di un moderno aereo in volo con uno sfondo di nuvole" l'algoritmo mi capisce e mi fornisce l'immagine richiesta.

Attenzione e molto importante: dobbiamo distinguere tra due tipi fondamentali di algoritmi: quello che agendo su un'immensa banca dati mi tira fuori tra le tante poesie ed immagini archiviate, quelle che più si avvicinano alla mia richiesta.

L'altro, più sofisticato (AGI), che qui utilizzeremo, crea dal nulla poesia ed immagine partendo dal "senso" della richiesta; in questo caso la poesia e l'immagine, una volta create, sono uniche e non sono presenti nella memoria del sistema; ripetendo la stessa richiesta il sistema fornirà risposte sempre diverse.

Anticipo che con la produzione di oggetti testuali e immagini sfruttando l'AGI nasce la grande possibilità di considerarli nostri esclusivi e quindi di poterli gestire come appartenenti al nostro diritto d'autore.

DIRITTO D'AUTORE

La questione dei diritti d'autore per i testi e le immagini generate da AI di tipo AGI, cioè proprio quelle che qui trattiamo, è un campo emergente che coinvolge sia la legge sul diritto d'autore sia la proprietà intellettuale.

In generale, quando si crea un'opera d'arte originale, inclusa la scrittura di un testo che viene utilizzato per generare una poesia o un'immagine attraverso un servizio AI, i diritti d'autore possono essere attribuiti a chi ha formulato il testo per crearli, a meno che i termini di servizio della piattaforma che si utilizza non lo vieti in maniera specifica ed evidente.

Nel caso che qui trattiamo e che si riferisce alla piattaforma di OpenAI, questo divieto non viene esplicitamente espresso nei suoi documenti pubblici e la piattaforma specifica esattamente quanto segue:

"La piattaforma OpenAI stabilisce delle linee guida e delle politiche per l'uso dei suoi strumenti, compresi i regolamenti sui diritti d'autore. Gli utenti sono responsabili per come usano gli output generati dagli strumenti OpenAI e devono assicurarsi di non violare i diritti d'autore di terzi. Ad esempio, la creazione di immagini che imitano lo stile di artisti moderni o l'uso di personaggi protetti da copyright nei prompt per la generazione di immagini è proibita.

Per quanto riguarda i testi o le immagini creati dalla piattaforma, gli utenti possono generalmente utilizzarli liberamente, ma ci sono delle eccezioni e delle specifiche condizioni che possono applicarsi. È importante leggere e comprendere i termini di servizio di OpenAI e le politiche sulla proprietà intellettuale per assicurarsi di conformarsi alle regole".

Aggiungo che per casi specifici gli autori possono interrogare la piattaforma stessa senza alcun problema.

In tutti gli esempi pubblicati in questo libro ho attentamente utilizzato miei testi di tipo generale e che non chiedevano alla piattaforma di imitare qualcuno o qualcosa, ma descrivo ciò che desideravo che mi creasse come ad esempio: "crea l'immagine di una città con abitanti in movimento del periodo dell'antica Roma", oppure: "crea l'immagine di un'astronave che si avvicina alla galassia Andromeda in cui si vedono i piloti che guardano in avanti la galassia" e così via.

Queste descrizioni sono generiche e l'immagine risultante posso ritenerla di mia proprietà. Spesso ho provato a ripetere la stessa identica domanda alla piattaforma e l'immagine risultava sempre diversa, per cui la prima creatami potevo considerarla "unica" e irripetibile., oltre che mia.

Questo significa anche che nessun lettore potrà riprodurre le immagini utilizzando i miei testi attraverso la piattaforma anche se pubblico i testi che ho utilizzato per generarle.

ALCUNI UTILI CONSIGLI

Essendo un utilizzatore fin da quando l'accesso a queste risorse è diventato pubblico, ho potuto constatarne le grandi possibilità, ma anche i rischi.

Posso con certezza affermare che per sfruttare al meglio mezzi del genere occorra avere una certa dimestichezza con l'argomento che si vuol condividere con l'AI. Mi spiego, a volte ho avuto vere e proprie "risse" con l'AI su argomenti di cui sono particolarmente esperto perché alla mia domanda tecnica la risposta era palesemente sbagliata, di ordini di grandezza. Io correggevo e l'AI mi ringraziava.

Mi è capitato raramente, ma devo proprio dire che, quando l'AI sbaglia, gli errori sono enormi, per cui attenti! L'uso di questi strumenti è stupendo per chi conosce l'argomento, ma possono essere fatali per chi è totalmente all'oscuro di ciò che chiede alla piattaforma. Mi verrebbe voglia di dire che stiamo parlando di strumenti fantastici per gli esperti, ma pericolosi per chi nulla conosce o desidera creare complete opere inventate.

Per farmi capire con un esempio semplice mi riferirò alla mia richiesta di un'immagine che troverete in questo libro. Ho richiesto alla piattaforma di "Creare un aereo passeggeri jet a tre motori della fine del secolo scorso tipo DC10" ed ecco cosa mi ha creato:

Ho chiesto alla piattaforma in tutti i modi possibili di togliere i motori in più e alla fine la piattaforma si è arrabbiata e mi ha scritto di utilizzare qualche software e di arrangiarmi! Ed infatti l'immagine che troverete in questo libro ha i due motori di troppo cancellati con Photoshop.

Naturalmente nel caso specifico sapevo bene che quell'aereo, creato su mia richiesta, non è mai esistito.

In un altro caso di tipo scientifico volevo verificare quanto avevo calcolato in un mio precedente libro sui cambiamenti climatici. Ho chiesto alla piattaforma di calcolarmi di quanto si sarebbero alzati gli oceani se tutti i ghiacci si fossero sciolti.

Ne è seguita una sequela di calcoli anche complessi sull'entità dei ghiacci ecc. ecc. io continuavo a correggere la mia domanda perché la piattaforma aggiustasse il tiro non tenendo conto di alcuni fattori. Anche qui alla fine ho dovuto dare forfait perché non riuscivo a spiegare all'AI che, se gli oceani si alzano, si allarga anche la superficie degli oceani e di questo occorreva tenerne conto ed altro, e questo è stato il risultato testuale:

"Mio commento: Se questa è la profondità media sciogliendosi tutti i ghiacci gli oceani si alzerebbero meno di 58 metri. RISPOSTA: hai ragione, mi scuso per l'incongruenza nei dati forniti in risposta alla tua domanda precedente. La profondità media degli oceani è effettivamente di circa 3.800 metri (12.467 piedi), e se tutto il ghiaccio sulla Terra si sciogliesse, l'aumento del livello del mare sarebbe significativamente inferiore a 58 metri.

Come menzionato in precedenza, se tutto il ghiaccio presente sulla Terra si sciogliesse, l'aumento del livello del mare sarebbe di circa 66-67 metri. Questa stima tiene conto di tutto il ghiaccio presente nelle calotte polari dell'Antartide e della Groenlandia, nei ghiacciai delle montagne, nell'Artico e in altre regioni.

Quindi, se il livello del mare aumentasse di 66-67 metri, non influenzerebbe direttamente la profondità media degli oceani, poiché la profondità media è una misura della profondità dall'attuale livello del mare fino al fondo degli oceani. In altre parole, la profondità media rimarrebbe sostanzialmente la stessa, ma la superficie dell'oceano si estenderebbe ulteriormente a causa del maggiore volume d'acqua.

Grazie per la correzione, e spero che questa risposta sia più chiara."

Chiaramente anche la risposta finale è sbagliata!

Detto questo, assicuro il lettore che nel 99% dei casi sono riuscito ad ottenere quello che volevo, magari con un po' di pazienza. Dove invece ho cannato alla grande è quando ho chiesto di scrivermi una poesia per mia moglie: ho dato alcune informazioni, l'ho stampata e gliel'ho consegnata ... c'è mancato poco che me la facesse ingoiare! Quindi, cari lettori, occhio a non disumanizzare troppo i vostri rapporti sfruttando l'AI!

I prodotti di OpenAI

Il modo migliore per conoscerli è chiederlo alla stessa piattaforma ed ecco l'ultima risposta:

"Al momento del mio ultimo aggiornamento nel gennaio 2022, OpenAI aveva diverse piattaforme e prodotti. Tuttavia, è possibile che siano stati introdotti nuovi prodotti o cambiamenti dalla mia ultima conoscenza. Le principali piattaforme offerte da OpenAI includevano:

GPT-3: Il modello GPT-3 (Generative Pre-trained Transformer 3) è uno dei prodotti principali di OpenAI ed è un modello di linguaggio altamente avanzato che può essere utilizzato per generare testo, rispondere a domande e molto altro ancora.

GPT-3 API: OpenAI ha reso disponibile un'API (Application Programming Interface) per GPT-3, consentendo agli sviluppatori di integrare GPT-3 nelle loro applicazioni e servizi.

OpenAI Codex: Questo prodotto è stato sviluppato in collaborazione con GitHub ed è basato su GPT-3. OpenAI Codex è un motore di generazione di codice che può aiutare gli sviluppatori a scrivere codice in vari linguaggi di programmazione.

DALL-E: DALL-E è un modello di intelligenza artificiale sviluppato da OpenAI che può generare immagini a partire da descrizioni testuali. Può creare immagini sorprendenti basate su concetti descritti in parole.

CLIP: CLIP è un altro modello di OpenAI che è stato addestrato per comprendere il contenuto visuale e testuale in modo congiunto. Può essere utilizzato per una vasta gamma di applicazioni legate all'immagine e al testo, come il riconoscimento di immagini e il recupero di testo.

OpenAI GYM: OpenAI GYM è una piattaforma per la ricerca in intelligenza artificiale che offre una suite di ambienti di addestramento per sviluppare e testare algoritmi di apprendimento automatico e di apprendimento rinforzato.

Diverse altre risorse e strumenti: OpenAI potrebbe avere introdotto ulteriori prodotti o risorse dalla mia ultima conoscenza. È consigliabile visitare il sito web ufficiale di OpenAI per le informazioni più aggiornate sulle loro offerte.

Tieni presente che le informazioni possono essere cambiate nel tempo; quindi, è sempre bene verificare il sito web di OpenAI o altre fonti aggiornate per le informazioni più recenti sui loro prodotti e servizi."

Immagini create con la piattaforma

Le immagini che seguono sono state create con DALL-E di Open IA per una visione completa da parte del lettore di quanto, con le opportune richieste anche fantastiche, sia possibile oggi ottenere con simili strumenti.

Sono immagini raccolte in un paio di mesi e di queste, alcune, faranno parte di un mio futuro libro.

Nota: le didascalie riportate in ogni immagine sono il testo che DALL-E utilizza per generare l'immagine e che è ricavate dalle mie richieste in italiano; non è quindi la traduzione del mio testo ma la sua interpretazione in inglese.

Il sistema accetta le richieste in qualsiasi lingua e vengono trasformate e quindi tradotte nel testo inglese che l'Intelligenza Artificiale è in grado di comprendere per creare l'immagine.

Il lettore potrà verificare come la maggior parte delle immagini siano meravigliose, alcune imperfette ed altre assurde ... e deve quindi intervenire l'intelligenza umana per scegliere e correggere.

Evoluzione dell'uomo

A group of early hominids walking upright through a prehistoric landscape. The scene captures the pivotal moment in human evolution with a clear sky overhead. These early hominids, possibly Australopithecus or early Homo species, are characterized by their bipedal gait, distinguishing them from their quadrupedal ancestors.

Scene inside a dimly lit cave where a group of Neanderthals are gathered around a fire. The flames illuminate their features and cast shadows on the wall, creating a captivating display of dancing light and darkness as they huddle together, their faces reflecting both the warmth of the fire and the intrigue of their shared ancestral stories.

A scene set during the Upper Paleolithic period showing a group of Homo sapiens in front of a cave entrance. They are engaged in the meticulous task of working stones, skillfully chipping away at flint and obsidian, their expert hands shaping these materials into tools and weapons that would be crucial for their survival in a harsh and unforgiving prehistoric world.

A dynamic and tense scene of a group of Homo sapiens in a confrontational stance, throwing stones at each other. The individuals are scattered across a rugged terrain, their faces contorted with anger and determination as they engage in a primitive, high-stakes conflict reminiscent of humanity's early struggles for survival.

A vivid scene of two Homo sapiens in the midst of a heated duel, using trees as part of their combat strategy. They are set in a dense forest environment, where ancient trees with gnarled roots and thick foliage provide both cover and obstacles. The combatants, faces streaked with war paint and eyes filled with determination, expertly employ the natural terrain to gain an advantage.

Scene of two groups of early humans facing each other in an aggressive stance, engaged in a stone-throwing battle in front of a cave entrance. They are depicted in a dramatic standoff, with furrowed brows and clenched fists, as they fling stones with force and precision toward their rivals.

Inside a warm, fire-lit cave, a couple of Homo sapiens are seen with their newborn. The father and mother, showing a mix of awe and affection, are sitting close to each other, their faces illuminated by the gentle flicker of the firelight. They cradle their precious infant in their arms, their eyes locked onto the tiny life they've brought into the world, their hearts filled with a profound sense of love and wonder.

A scene of a Homo sapiens hunter during the late Paleolithic era, poised with a bow and arrow, aiming at a deer in the distance. The setting is a lush, prehistoric landscape, with tall grasses and dense foliage that partially conceal the hunter as they patiently track their prey through the ancient wilderness. The tension in the air is palpable as the hunter carefully lines up the shot, relying on their skill and precision to secure sustenance for their tribe.

An Ice Age scene where a group of Homo sapiens are strategically positioned around a trap they have set in the ice. The trap is a large hole that has been meticulously carved through the thick ice of a frozen river or lake. The humans, bundled up in furs to protect against the frigid cold, are hidden in camouflaged positions, waiting for a herd of mammoths or other large prehistoric creatures to approach.

A striking scene from the Ice Age, depicting a group of Homo sapiens surrounding a large hole in the ice where a mammoth has tragically fallen. The humans, their faces a mix of awe and trepidation, have gathered around this unexpected bounty, realizing the importance of their find for their survival. With makeshift tools and cooperative effort, they are preparing to harvest meat, ivory, and other valuable resources from the massive creature.

A vivid image of an Ice Age scenario with a group of Homo sapiens gathered around a significant hole in the ice. In the scene, a mammoth has been captured and immobilized, its massive form partially submerged in the icy water. The humans, bundled in fur clothing and with determined expressions, work together to secure their impressive prize.

A detailed image of a mammoth trapped in a deep ice hole during the Ice Age. The mammoth is visibly large, with its shaggy fur and immense tusks, and it stands with its front legs in the icy pit while the rest of its massive body is outside. Its trunk is raised in a mix of distress and curiosity as it surveys the humans gathered around the hole.

An evocative image capturing a group of Homo sapiens in the midst of a mammoth hunt in a glacial landscape. The hunters are spread out, using primitive spears and teamwork to approach the massive creature. The icy terrain stretches out before them, with towering glaciers and snow-covered mountains forming a majestic backdrop to the hunt.

A vivid depiction of an Ice Age hunting scene where a mammoth is fully enclosed within a deep snow pit. Above the pit, three Homo sapiens hunters are strategically positioned on the snowy terrain. They hold long, sturdy spears tipped with sharpened stones, and their faces are etched with concentration and determination.

An intense scene of two Homo sapiens engaged in combat using wooden clubs during the prehistoric era. They stand on a rough, grassy terrain with the fiery hues of a setting sun casting long shadows across the landscape. The combatants, their faces contorted with determination, swing their primitive weapons with precision and force, locked in a fierce struggle for dominance.

An intense scene of two Homo sapiens engaged in combat using wooden clubs during the prehistoric era. They stand on a rough, grassy terrain with the wind sweeping through the tall, swaying grasses. The combatants, their brows furrowed with determination, wield their primitive clubs with skill and ferocity, their bodies painted with tribal markings that accentuate the primal nature of their struggle.

Here is the image of the historical battle scene from the early modern period, with soldiers engaged in combat using early matchlock guns and wearing period armor.

A historical battle scene with men engaged in combat using very primitive firearms, such as hand cannons and early matchlock muskets. The setting is a battlefield shrouded in smoke and chaos, where the deafening roar of cannon fire and the acrid smell of gunpowder fill the air, creating a scene of intense and brutal warfare.

A dynamic medieval battle scene where soldiers with early firearms are attacking a castle. The castle stands formidable with its stone walls and battlements, its towering towers looming over the attackers like silent sentinels of a bygone era. The defenders, clad in armor and armed with bows, crossbows, and swords, stand resolute atop the walls, determined to repel the relentless assault from below. The clash of technology and tradition creates a gripping tableau of a changing world caught in the throes of history.

A vivid depiction of a medieval siege where the warriors are now facing towards a formidable stone castle. The castle stands majestic on a steep hill, its battlements lined with archers ready to defend against the advancing forces below. Siege engines prepare to launch another assault, with trebuchets and catapults poised to hurl stones against the thick walls. The sky is a dramatic canvas of brooding clouds, hinting at an impending storm that adds to the tension of the battlefield. Flags bearing the emblems of both the besieging army and the defending lords flutter in the gusty winds, symbolizing the clash of wills and steel that is about to unfold.

Il trasporto nel passato

A prehistoric scene depicting early humans moving a giant stone by rolling it over a series of logs. The stone is massive, possibly a future megalith, and the primitive humans, with sinewy muscles and simple tools, work together in a coordinated effort to slowly inch the colossal rock across the rugged terrain. The scene is a testament to their ingenuity and strength, as they laboriously transport the stone to its intended location, where it will eventually become a monumental part of their ancient culture and history.

A dynamic prehistoric scene showing several humans with different physical characteristics working together to transport a huge boulder. The group is a diverse assembly of early humans, each with unique physical attributes suited to the task at hand. Some individuals are strong and muscular, while others possess agility and endurance. They collaborate seamlessly, using their varied skills to maneuver the massive boulder across challenging terrain. This cooperative effort highlights the strength of their social bonds and the collective ingenuity that allowed early humans to achieve remarkable feats, even in the distant past.

An ancient Egyptian landscape, bustling with activity as people transport goods along the Nile River. Several wooden boats with large sails, typical of ancient Egyptian vessels, glide gracefully on the tranquil waters. The riverbanks are lined with bustling markets where merchants and traders barter for various goods, while farmers tend to their fields adjacent to the fertile riverbanks, taking advantage of the annual Nile floods to grow crops. In the distance, the majestic pyramids and temples of Egypt rise against the horizon, casting imposing shadows on the landscape. This scene captures the essence of ancient Egypt, a civilization that thrived along the banks of the Nile for millennia, thanks to the life-giving waters of this mighty river.

A lively scene from the Egyptian period showcasing horse-drawn chariots. The setting is an expansive desert landscape with the Nile River flowing in the background. In the foreground, powerful horses strain against their harnesses, pulling ornate chariots with skilled charioteers gripping the reins. These chariots are adorned with intricate designs and symbols, reflecting the grandeur of ancient Egyptian craftsmanship.

An ancient Roman biga, a lightweight chariot drawn by two horses side by side, moving swiftly on a dusty racetrack. The horses, adorned with decorative harnesses and colorful plumage, exert themselves as they gallop in perfect synchrony, their hooves pounding the earth rhythmically. The charioteer, clad in a vibrant tunic and gripping the reins with determination, leans forward in a display of skill and daring, guiding the chariot with precision as the crowd in the grandstands roars with excitement. This exhilarating scene captures the essence of Roman chariot racing, a beloved and thrilling sport in ancient times.

A medieval carriage on a cobblestone road in a European village. The carriage is wooden with metal reinforcement, drawn by two robust horses with plumes adorning their harnesses. It creaks and rattles as it rolls along the uneven cobblestones, the sound echoing through the narrow streets lined with charming, timber-framed houses. The carriage's passengers, dressed in the fashion of the time, peer out from curtained windows, their journey through the picturesque medieval village a blend of rustic elegance and historical charm.

A sophisticated Renaissance carriage on a bustling street of an Italian city. The carriage is ornate, with a curved roof and intricate patterns engraved into its polished wood. It is pulled by a team of beautifully groomed horses, their luxurious harnesses adorned with gold accents and colorful ribbons. As the carriage navigates the cobblestone streets, pedestrians stop to admire its elegance and craftsmanship, while the occupants inside, dressed in luxurious Renaissance attire, exude an air of refinement and opulence. This scene encapsulates the grandeur and cultural richness of the Renaissance era in Italy.

A sophisticated late 1800s steam-powered automobile, featuring a high, curved roof and large wooden wheels. The vehicle is adorned with ornate metalwork, including brass fittings and intricate scrollwork on the headlamps and grille. Its polished wooden body gleams in the sunlight, and a small plume of steam can be seen escaping from the rear exhaust. The driver's seat is elevated, with a leather-upholstered bench, and a brass horn is mounted to the side for signaling. Elegantly dressed passengers could be seen inside, enjoying the novelty and luxury of early motorized transport.

Residenze umane nella storia

Visualize a group of prehistoric stilt houses, set in a marshy region with a network of waterways. The dwellings are perched atop tall wooden stakes, their foundations rising above the water's surface, creating a captivating and surreal landscape. Surrounding the stilt houses, the water reflects the sky's hues, mirroring the vibrant colors of the natural world. These ancient homes, suspended above the water, served as safe havens for early human communities, providing protection from both rising waters and potential predators, while also facilitating transportation and access to vital resources in the marshy environment.

Visualize an early human encampment from the dawn of civilization. The scene is set in a lush landscape with primitive dwellings made of sticks, straw, and animal hides. These simple, makeshift structures dot the landscape, nestled amidst the natural beauty of rolling hills and meandering rivers. In the heart of the encampment, a communal fire burns, its flickering flames casting a warm, inviting glow,

Imagine the ancient city of Babylon during the Assyrian period, complete with the fabled Hanging Gardens. The city is a marvel of Mesopotamian architecture, with towering ziggurats and intricate palace complexes adorned with colorful tiles and ornate carvings. Amidst this urban splendor, the Hanging Gardens, one of the Seven Wonders of the Ancient World, stand as a lush oasis in the heart of the metropolis. Vibrant plants and trees cascade down tiered terraces, creating a breathtaking spectacle of greenery and beauty.

Visualize a bustling ancient Egyptian city during the time of the pharaohs. The city is alive with vibrant market streets where traders sell linen, jewelry, and exotic spices from lands near and far. The aroma of incense wafts through the air as street vendors offer tantalizing foods and perfumes to passersby. Towering above the busy streets are grand temples and statues dedicated to the gods, their majestic presence a testament to the spiritual significance of the city.

Imagine the splendor of ancient Rome with a panoramic view that includes the sprawling Roman Forum and distant Palatine Hill. In the foreground, the Forum is a bustling center of activity, with Roman citizens and senators clad in togas, going about their daily business amidst the grandeur of ancient monuments and temples. The air is filled with the sound of discussion and debate as the heart of the Roman Republic pulses with life.

Create a detailed image of a medieval cityscape at the peak of the feudal era. The view overlooks a bustling medieval town with a fortified stone castle standing proudly at its center. The town is a labyrinth of narrow cobblestone streets, winding past half-timbered houses with thatched roofs, and market stalls where merchants peddle their wares.

Create a detailed image of a medieval cityscape at the peak of the feudal era. The view overlooks a bustling medieval town with a fortified stone castle standing proudly at its center. The town is a labyrinth of narrow cobblestone streets, winding past half-timbered houses with thatched roofs, and market stalls where merchants peddle their wares.

An ultra-modern cityscape during the day, showcasing a clear sky and the sun shining over an array of futuristic skyscrapers. These structures have a sleek and minimalist design, characterized by smooth, reflective surfaces of glass and steel that glisten in the sunlight. The buildings reach towards the heavens, their geometric shapes forming a stunning contrast against the azure backdrop.

A futuristic cityscape with towering skyscrapers, featuring a skyline dotted with advanced architectural designs. The buildings are equipped with sleek, curved exteriors that seem to blend seamlessly into the sky, and their surfaces shimmer with an ever-changing array of colors and patterns thanks to advanced LED lighting.

Imagine a cityscape years ahead of our time, bustling with people from all walks of life, equipped with state-of-the-art wearables and devices. In the sky, drones zip back and forth, some delivering goods while others project interactive advertisements. The buildings are tall and made of smart materials that change color with the weather. The roads are filled with autonomous vehicles moving in harmony, while above, a network of skybridges connects multiple levels of the city, allowing pedestrians to navigate the urban expanse with ease. Green spaces float between structures, purifying the air and bringing nature into this high-tech urban environment.

Lo sviluppo dell'automobile

A late 19th-century automobile on a cobblestone street in a Victorian-era city. The automobile is an intricate steam-powered vehicle with a large boiler at its front, billowing white plumes of steam as it chugs along. Its polished brass and wooden accents gleam in the soft glow of gas street lamps lining the road.

An authentic representation of an early 1900s automobile. The vehicle is characterized by its vintage charm, featuring large spoked wheels, a curved ornate radiator grille, and a long, sleek hood. The body is adorned with hand-painted pinstripes and elegant detailing, showcasing the craftsmanship of the era.

A classic scene with a female figure seated at the wheel of an early 1900s automobile. She is dressed in period-appropriate attire, wearing a long, flowing dress with a high neckline and a wide-brimmed hat adorned with ribbons. Her gloved hands confidently grip the large, wooden steering wheel.

A classic early 20th-century motor carriage, reminiscent of the transitional period from horse-drawn carriages to automobiles. The vehicle has large spoked wheels and a vintage, boxy design that harkens back to the days of horse-drawn coaches. The wooden carriage body is painted in a rich, glossy color, and its exposed brass accents gleam in the soft sunlight.

A modern car from the period of 1900-1910, displaying the transitional automotive design of the era. The car, sleek for its time, has a glossy black body with polished brass trimmings, including the headlamps and radiator grille. Its large spoked wheels are fitted with solid rubber tires, a sign of progress from the earlier days of horse-drawn carriages.

An image of an old modern-style sedan car influenced by the iconic designs of early Ford models. The vehicle has a classic black body with a glossy finish that exudes timeless sophistication. Its exterior features clean lines and understated elegance, reminiscent of the Model T and other early Ford classics.

A detailed scene set in the 1930s featuring two people inside a classic sedan car. The car has the distinctive design of the era, with a long, curved body that exudes Art Deco elegance. Its glossy exterior is a deep, luxurious color, and the chrome accents on the grille and bumpers gleam in the soft, golden afternoon sunlight.

A depiction of an old yet modern-looking sedan car, illustrating the aesthetic of timeless automotive design. The vehicle has a sleek, black body with clean, elegant lines that give it a sense of understated sophistication. The chrome accents on the door handles, grille, and trim are polished to a high shine, adding a touch of luxury to the overall design.

A depiction of an old yet modern-looking sedan car, illustrating the aesthetic of timeless automotive design. The vehicle has a sleek, black body with clean, elegant lines that give it a sense of understated sophistication. The chrome accents on the door handles, grille, and trim are polished to a high shine, adding a touch of luxury to the overall design.

An image of a classic modern-style sedan car, reminiscent of the luxury and grandeur associated with an early Rolls-Royce. The vehicle features a two-tone color scheme, with a deep, glossy black body and contrasting ivory accents. Its imposing grille is adorned with a gleaming chrome Spirit of Ecstasy emblem, and the classic wire-spoke wheels are polished to perfection. artistic representation of an old modern-style sedan car, inspired by the luxurious and sporty design of an early Bugatti. The car boasts a sleek, aerodynamic body in a vibrant racing blue color, with contrasting polished aluminum accents that catch the light. Its curvaceous lines and low-slung profile evoke a sense of both elegance and speed.

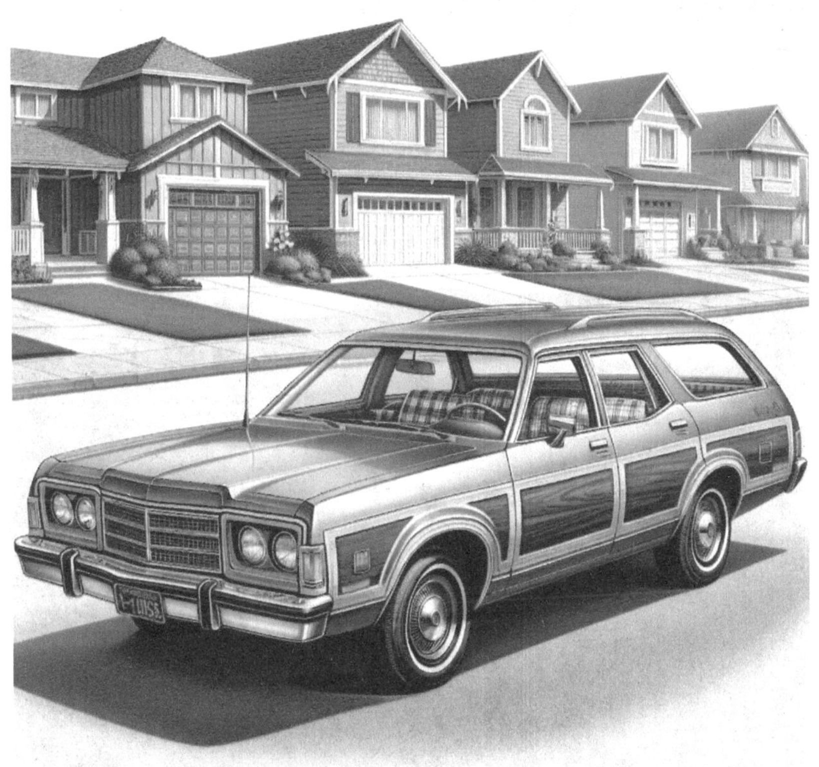

A depiction of a 1970s four-door sedan car, showcasing the era's distinct automotive style. The car is finished in a classic avocado green with a faux wood-grain panel running along its sides. Its boxy design is both functional and practical, reflecting the aesthetics of the time.

An image of a 1970s sedan car that exemplifies the style and trends of the era. The car has a mustard yellow body with chrome trim and features a characteristic vinyl roof in a contrasting darker shade, adding a touch of luxury to its appearance. Its body is sleek and angular, with a long hood and distinctive rectangular headlights.

A classic 1960s four-door sedan car, embodying the automotive design ethos of the era. The vehicle is finished in a glossy cherry red with a contrasting white roof, a popular two-tone style of the time. Its body boasts sleek, flowing lines, and the chrome accents, including the bumpers, grille, and side trim, are polished to a high shine.

A depiction of a classic 1980s sedan car, reflecting the design trends of the era. The car has an angular design with a distinct wedge shape, featuring a glossy midnight blue body. Its boxy silhouette is marked by sharp lines and geometric patterns, which were characteristic of automotive design during the 1980s.

An image of a 2010s sedan car that showcases the sleek aesthetics and innovation of the decade. The car displays a contoured body with smooth lines, featuring a metallic silver finish that glistens in the sunlight. Its design is marked by a blend of aerodynamic efficiency and modern sophistication, a hallmark of automotive styling in the 2010s.Figure 1

An image of a 2010s luxury sedan car, embodying the advanced engineering and design language of the era. The car features a sophisticated silhouette with a low, sweeping roofline and a metallic pearl white finish that radiates elegance. Its sleek lines and aerodynamic contours are designed to maximize both performance and fuel efficiency.

An image of a modern sedan car with the sophistication and sleek design characteristic of a contemporary Mercedes-Benz. The vehicle features a smooth, aerodynamic silhouette with a glossy finish, reflecting the latest in automotive innovation. Its grille boasts the iconic emblem, flanked by LED headlights with a distinctive, angular shape. The car's profile is low and elegant, with a sloping roofline that merges seamlessly into the rear deck. Chrome accents highlight the windows and door handles, adding a touch of luxury to the deep black or metallic silver paint. This sedan is parked on a pristine driveway, the embodiment of modern engineering and design finesse, ready to glide along the city streets or open highway.

An image of a 2020s electric sedan car, epitomizing the cutting-edge design and eco-friendly technology of the decade. The car has an aerodynamic silhouette with a glossy, pearlescent white finish that emphasizes its modernity and environmental consciousness. Its sleek lines and smooth curves are not only aesthetically pleasing but also designed for maximum. efficiency.

An image of a modern 2020s electric sedan, representing the sleek design and innovative technology of today's electric vehicles. The car has a smooth, aerodynamic silhouette with a stunning pearlescent silver finish that accentuates its contemporary style and environmental consciousness. Its clean lines and well-proportioned body are not only aesthetically pleasing but also optimized for maximum efficiency.

An image of a 2020s electric sedan car, featuring the clean, futuristic lines that define electric vehicle design. The car is showcased in a sleek graphite gray finish, emphasizing its modernity and commitment to sustainability. Its sculpted body and aerodynamic profile not only enhance its aesthetic appeal but also contribute to its energy efficiency.

Lo sviluppo dei treni

An illustration of the very first steam locomotive from the early 1800s, showcasing its pioneering design. The locomotive should depict a basic, rudimentary structure with a large, cylindrical boiler at its center. The boiler is mounted on a simple frame with spoked wheels, and a tall, slender smokestack rises from the front of the boiler. The locomotive's design is minimalist, with exposed gears and mechanisms that power the locomotion

A panorama showcasing a chronological lineup of steam locomotives from the early 1800s to slightly more recent designs. On the far left, the very first steam locomotive, resembling a basic, rudimentary structure with a large, cylindrical boiler and exposed gears, represents the early 1800s. On the right, the locomotives gradually evolve in design and sophistication.

A depiction of a very ancient steam locomotive, even older in design than the previous models from the 1800s. This locomotive resembles the earliest prototypes, featuring an extremely basic, almost primitive structure. The locomotive is composed of a rudimentary boiler mounted on a wooden frame with large, spoked wheels. It lacks many of the refinements seen in later designs and has exposed gears and mechanisms.

An image of an ancient steam locomotive, with intricate detailing and rustic charm, chugging along a vintage railroad track amidst a landscape reminiscent of the 19th century. The locomotive is a beautifully restored piece of history, adorned with ornate brass accents and polished woodwork. Its large spoked wheels and elegant smokestack capture the essence of early locomotive design. The body of the locomotive is painted in rich, deep colors, and its headlamp casts a warm, inviting glow.2

A vintage steam locomotive from the 19th century, featuring a classic design with a prominent cowcatcher, large spoked wheels, and an ornate, curved smokestack. The locomotive exudes a sense of historical charm and elegance, harkening back to the golden age of steam railroads. The cowcatcher at the front of the locomotive is a distinctive feature, designed to clear debris from the tracks. Its brass detailing and polished woodwork add a touch of luxury to the design. The wheels are robust and sturdy, built for the rugged terrain of the time.

An old-fashioned steam locomotive, with a large smokestack belching white smoke, rugged iron wheels, and a classic black body with silver accents. The locomotive is a symbol of the bygone era of steam-powered rail travel, featuring a timeless and iconic design. The smokestack releases billowing white clouds of steam as the locomotive powers forward on iron tracks. Its wheels are sturdy and built for traversing long distances, and the black body of the locomotive is adorned with polished silver detailing and accents.

A sleek, ultra-modern high-speed train, cutting-edge in design, gliding along futuristic magnetic levitation (maglev) tracks. The train features a streamlined, aerodynamic body with a glossy silver finish, exuding a sense of speed and technological advancement. The maglev technology allows the train to hover effortlessly above the tracks, eliminating friction and enabling it to achieve incredible speeds. The train's large, panoramic windows offer passengers breathtaking views of the landscape as they travel at high velocity.

A detailed image of Leonardo da Vinci's flying machine, as conceptualized in his sketches. The apparatus features large, bat-like wings made of lightweight materials such as wood and fabric. These wings have a wingspan that stretches gracefully to capture the air, resembling the wings of a bird or a bat. The flying machine's body is composed of a wooden framework, intricately designed to support the wings and the pilot's position. The pilot, dressed in Renaissance-era clothing, stands inside the frame, grasping the handles attached to the wings.

An image of one of the earliest hot air balloons, also known as a Montgolfier, with its characteristic spherical shape. The balloon is made of a canvas-like material, and it is tethered to the ground with ropes. A small wooden basket, or gondola, is suspended beneath the balloon, carrying a few adventurous passengers.

At the base of the balloon, a fire burns in a specially designed brazier. The flames heat the air inside the balloon, causing it to expand and become ighter than the surrounding air. This lift from the hot air allows the balloon to rise gracefully into the sky.

An image of a modern hot air balloon, featuring a large and teardrop-shaped envelope made from durable nylon or polyester fabric. The balloon's envelope is a canvas for vibrant, colorful artwork, creating a stunning visual display against the backdrop of the sky. The balloon's design may include intricate patterns, abstract art, or themed illustrations that wrap around the envelope's surface. Bright shades of red, orange, yellow, blue, and green create a dynamic and eye-catching appearance. The balloon is inflated and stands tall, ready to take to the skies.

An illustration of one of the first dirigibles, showcasing its classic elongated shape and rigid structure. The dirigible features a slender, cylindrical body with an internal frame, likely made from lightweight yet sturdy materials such as aluminum or steel. The body of the dirigible is covered with a fabric envelope, creating a streamlined and aerodynamic appearance. The envelope is painted in traditional colors, often with the name or branding of the airship displayed prominently on its side.

Visualize the iconic Zeppelin dirigible, with its massive and elongated balloon-like body made of duralumin framework and covered in fabric. The gondola hangs beneath, housing the crew and passengers, its windows offering panoramic views of the landscapes below. Propellers attached to the sides of the gondola propel the vessel through the air, and the rear of the Zeppelin is adorned with rudders and elevators for steering. Floating gently in the sky, this marvel of early aviation cruises above the clouds, a symbol of human ingenuity and the golden age of airship travel.Figure 3

An image of the Wright brothers' first airplane, the Wright Flyer. The aircraft is depicted with its distinctive double-wing structure and wooden frame. The wings have a biplane configuration, with the lower wing slightly forward of the upper wing. The aircraft is positioned on a sandy beach, as it was during the Wright brothers' historic first powered flight at Kitty Hawk, North Carolina, in 1903. The Wright Flyer is mounted on a wooden rail, ready for takeoff.

An image of a World War I biplane, characterized by its double-wing structure and open cockpit. The aircraft features a biplane configuration, with two sets of wings arranged one above the other, connected by a system of struts and wires. The aircraft's body is constructed from a combination of wood and fabric, giving it a lightweight yet sturdy appearance. At the front of the biplane, there is a large, wooden propeller that is ready to start spinning. The engine, likely a radial engine, is concealed within the fuselage, providing the necessary power for flight.

An image of a 1930s airmail plane, featuring the sleek, single-wing design typical of early air mail aircraft. The plane is shown in flight against a backdrop of blue skies and puffy white clouds. The aircraft has a streamlined, elongated fuselage with a single, high-mounted wing that spans the width of the plane. The wings are thin and swept back, emphasizing speed and efficiency. The body of the plane is painted in a classic silver or aluminum finish, reflecting the aesthetics of the era.

Volare moderno

An illustration of one of the first passenger airplanes from the 1930s. The plane features a polished metal body with a high-wing design for stability and ample cabin space. The aircraft is showcased against a blue sky with fluffy white clouds. The airplane's high-wing configuration provides stability and ample space for passengers in the cabin below. The polished metal exterior reflects the sunlight, giving the plane a sleek and elegant appearance.

A nostalgic 1940s scene with two people looking out of the window of an airplane. The interior of the aircraft features the characteristic design of the era, with a warm and inviting atmosphere. The airplane's cabin is adorned with polished wooden paneling, giving it a cozy and elegant ambiance. The seats are upholstered in rich, deep colors and plush fabrics, providing comfort to passengers during their journey. The windows are framed in chrome accents, adding a touch of sophistication to the cabin's decor.

An illustration of one of the first passenger airplanes from the 1940s. The aircraft features a sleek aluminum body with a low-wing design, which was typical of the period and contributed to improved stability and aerodynamics. The airplane is shown against a blue sky, with its polished metal exterior reflecting the sunlight. The cabin windows are visible, and passengers can be seen through them, seated in rows of seats. The interior is designed with a blend of functionality and comfort, providing a pleasant experience for travelers.

An image of one of the first four-engine passenger airplanes from the 1940s. The aircraft boasts a significant and robust fuselage designed to accommodate a larger number of passengers, marking a significant advancement in aviation during that era. The airplane's exterior is painted in classic colors of the time, with a polished aluminum finish and the airline's logo on the tail. The four powerful engines are mounted on the wings, enhancing the aircraft's performance and range.

An image of a Douglas DC-6 passenger airplane from the 1950s, prominently featuring the aircraft's four engines. The DC-6 is captured in all its classic glory against a blue sky with scattered clouds.
The aircraft's exterior is adorned with the livery of the airline, with polished aluminum surfaces and the airline's logo on the tail. The four engines, each with a distinctive propeller, are mounted on the wings, and they convey a sense of power and reliability.

An illustration of a Boeing 747 jumbo jet, an iconic airplane that emerged in the late 1960s. The image prominently displays the 747's distinctive hump on the upper deck, which is a defining feature of this aircraft. The Boeing 747 is shown against a clear blue sky, emphasizing its grandeur and size. Its exterior is painted in the livery of the airline, with the airline's logo and branding on the tail. The aircraft's four powerful engines are visible under its wings.

An image of a 1960s jet airplane, showcasing the sleek design and technological advancements of the era. The aircraft features a smooth, aerodynamic body that exemplifies the innovative and futuristic aesthetics of the time. The jet is depicted against a blue sky with a few wispy clouds, highlighting its streamlined and elegant silhouette. Its exterior is painted in the livery of the airline, with polished metal accents and the airline's logo on the tail. The aircraft's engines, mounted beneath the wings, convey power and efficiency.

An image of a three-engine jet airplane, similar in design to the McDonnell Douglas DC-10. The aircraft is shown against a clear blue sky, highlighting its classic and distinctive features. The exterior of the airplane is painted in the livery of the airline, with polished metal accents and the airline's logo on the tail. The three powerful engines are prominently displayed, with one engine mounted under each wing and a third engine at the base of the vertical stabilizer on the tail.

An illustration of a Concorde supersonic passenger airliner in flight over the Atlantic Ocean. The sleek white aircraft, with its distinctive droop nose, is soaring gracefully against the backdrop of a deep blue sky and the vast expanse of the ocean. The Concorde's slender delta wing design is evident, emphasizing its supersonic capabilities. Its nose is lowered, showcasing the iconic feature that allowed for improved visibility during takeoff and landing.

An illustration of an Air France Concorde supersonic passenger jet flying over Paris. The iconic aircraft is depicted in its classic blue and white livery, soaring elegantly above the cityscape of Paris. The Concorde's slender delta wings and distinctive shape are on full display as it gracefully moves through the sky. The aircraft's nose is in its characteristic droop position, providing improved visibility during takeoff and landing.

Illustration of a futuristic suborbital aircraft designed for high-speed international travel, flying above Earth's atmosphere. The sleek, elongated vehicle is showcased against the backdrop of the curvature of the Earth below.The suborbital aircraft features a streamlined, aerodynamic design, with a metallic exterior and advanced propulsion systems. Its wings, if present, are positioned for efficient high-speed flight.

Figure 4 Illustration of a futuristic suborbital aircraft designed for high-speed international travel, flying above Earth's atmosphere. The sleek, elongated vehicle is showcased against the backdrop of the curvature of the Earth below. The suborbital aircraft features a streamlined, aerodynamic design, with a metallic exterior and advanced propulsion systems. Its wings, if present, are positioned for efficient high-speed flight. The aircraft travels.

Viaggiare nello spazio

Visualize a futuristic spacecraft landing on the Martian surface, with a backdrop of the red, rocky terrain characteristic of Mars. The spacecraft showcases advanced technology, with sleek, metallic surfaces and intricate details. The spacecraft's landing gear gently touches down on the Martian soil, creating a small cloud of dust as it settles. The surface of Mars stretches out as far as the eye can see, with its iconic red hue and scattered rocky formations.

Visualize two futuristic spacecraft landing on the Martian surface. The scene showcases the red, rocky terrain characteristic of Mars, with two sleek and advanced spacecraft touching down simultaneously.
The Martian landscape stretches out beneath the spacecraft, with its distinctive red soil and scattered rocky formations. The spacecraft's landing gear gently makes contact with the Martian surface, stirring up a cloud of rust-colored dust as they settle in.

Envision a Martian landscape where two futuristic spacecraft have landed on the rocky, red terrain. In this scene, astronauts in modern space suits are stepping out of the spacecraft to explore the Martian surface.
The landscape is dominated by the rust-colored soil of Mars, with scattered boulders and rocks in the distance. The spacecraft have touched down gently, with their landing gear leaving imprints on the Martian surface. The spacecraft themselves are sleek and metallic, showcasing advanced technology and innovative design.

Visualize a top-down aerial perspective of the scene of two futuristic spacecraft landing on the Martian surface. The view captures the spacecraft in the midst of their descent, with the Martian landscape stretching out below. From this vantage point, you can see the two sleek spacecraft as they approach the red, rocky terrain of Mars. Their landing gear is deployed, and they are descending gracefully towards the Martian surface. Dust and debris are kicked up by their engines as they make their final descent.

above the Earth's atmosphere, where the sky transitions from the blue of the troposphere to the deep black of space.

Design of a futuristic passenger spacecraft in low Earth orbit. This advanced vessel boasts a smooth, aerodynamic shape with reflective surfaces and neon accents, symbolizing the sleek and high-tech aesthetics of future space travel. The spacecraft's exterior is constructed from advanced materials that provide both durability and style. Its metallic surfaces gleam in the sunlight, reflecting the Earth below and the stars above.

A futuristic interplanetary passenger spaceship cruising through the asteroid belt between Mars and Jupiter. The ship showcases a streamlined, silver hull with advanced design elements that optimize its journey through the asteroid-filled region. The spaceship's exterior is constructed from advanced materials that provide both durability and functionality. I

A futuristic interplanetary passenger spaceship cruising through the asteroid belt between Mars and Jupiter. The ship showcases a streamlined, silver hull with advanced design elements that optimize its journey through the asteroid-filled region. The spaceship's exterior is constructed from advanced materials that provide both durability and functionality. Its sleek, silver surfaces glisten in the sunlight, offering a striking contrast to the dark backdrop of space and the countless asteroids scattered throughout the asteroid belt.

A close-up view of the Moon from the perspective of a spaceship's front window. The spaceship itself is not visible in the frame; the focus is entirely on the Moon's cratered surface. Through the spaceship's expansive window, the Moon's rugged and barren landscape comes into view. The surface is covered in countless craters of varying sizes, each a testament to billions of years of meteorite impacts.

Two astronauts, one Black female and one Hispanic male, conducting a spacewalk to repair a spacecraft. They are equipped with high-tech space suits an advanced tools to ensure their safety and success on this important mission, showcasing the diversity and capabilities of NASA's talented astronaut corps.

A majestic image of a futuristic spaceship on approach to Saturn. The sleek, metallic spaceship is in the foreground, guided by two robots within its command center, while the magnificent rings of Saturn loom in the distance, creating a breathtaking scene that captures the wonder of interplanetary exploration. 🚀🪐 #SaturnAdventure #FuturisticExploration

A breathtaking image of a futuristic spaceship advancing towards the Andromeda Galaxy. The spaceship is designed with sleek lines and advanced technology, and it's soaring through the cosmos with grace and purpose. The spaceship's exterior gleams with polished surfaces that reflect the cosmic light around it. Its streamlined design, reminiscent of the finest spacecraft of the future, signifies its advanced propulsion and navigation systems.

An awe-inspiring image of a futuristic spaceship on its approach to a black hole. The vessel, guided by its robotic pilots, is captured in the foreground against the backdrop of the colossal, swirling black hole. The spaceship's exterior is a marvel of advanced engineering, featuring a sleek and sturdy design built to withstand the extreme gravitational forces and radiation near the black hole. Its metallic surface reflects the brilliant light and energy emitted by the accretion disk surrounding the black hole.

An intriguing image of a futuristic spaceship nearing Phobos, one of Mars' moons. The spaceship, a sophisticated craft with a polished metallic finish, is seen in close proximity to the irregularly shaped moon. The spaceship's exterior is a testament to advanced aerospace engineering, featuring a gleaming, silver surface that reflects the sunlight as it approaches Phobos. Its streamlined design and advanced propulsion systems demonstrate its capability to navigate the Martian system with precision.

A striking image of a futuristic spaceship with two robots in the cockpit as it approaches the Moon. The robots have a polished metallic design that glistens in the light, and they are seated in the cockpit, working together to guide the spaceship safely toward its lunar destination. The spaceship itself is a marvel of advanced technology, featuring a sleek and aerodynamic design that cuts through the vacuum of space with ease. Its exterior is constructed from advanced materials, and it boasts state-of-the-art propulsion systems for precision navigation.

An awe-inspiring image of a sleek, futuristic spaceship approaching Jupiter. The vessel, with its clean lines and illuminated panels, is piloted by two astronauts who are focused on navigating their advanced spacecraft through the cosmic expanse. The spaceship's exterior showcases cutting-edge design, featuring a streamlined, metallic surface that gleams as it approaches the gas giant. Its advanced propulsion systems and navigation instruments are a testament to human engineering at its finest.

An intriguing image of a futuristic spaceship nearing Phobos, one of Mars' moons. The spaceship, a sophisticated craft with a polished metallic finish, is seen in close proximity to the irregularly shaped moon. The spaceship's exterior is a testament to advanced aerospace engineering, featuring a gleaming, silver surface that reflects the sunlight as it approaches Phobos. Its streamlined design and advanced propulsion systems demonstrate its capability to navigate the Martian system with precision.

An atmospheric image of a futuristic spaceship approaching Deimos, the smaller moon of Mars. The spacecraft, featuring a streamlined design and operated by highly advanced technology, is seen gracefully closing in on the diminutive moon. The spaceship's exterior gleams with a polished metallic finish, reflecting the dim Martian sunlight as it approaches Deimos. Its sleek design is complemented by state-of-the-art propulsion systems that ensure a smooth and controlled approach.

A futuristic spaceship soaring through the cosmos, with sleek metallic surfaces reflecting the distant stars. The design is cutting-edge, featuring an elegant and streamlined silhouette that represents the pinnacle of aerospace engineering. The spaceship's exterior is adorned with a polished metallic finish that gleams in the cosmic light, creating a striking contrast against the backdrop of the vast, star-studded universe. Its advanced propulsion systems propel it effortlessly through the infinite expanse of space.

A futuristic spaceship soaring through the cosmos, with sleek metallic surfaces reflecting the distant stars. The design is cutting-edge, featuring an elegant and streamlined silhouette that represents the pinnacle of aerospace engineering. The spaceship's exterior is adorned with a polished metallic finish that gleams in the cosmic light, creating a striking contrast against the backdrop of the vast, star-studded universe. Its advanced propulsion systems propel it effortlessly through the infinite expanse of space.

Create an image of a futuristic spaceship en route to Jupiter. The spaceship is a marvel of cutting-edge design, featuring a sleek, streamlined hull that effortlessly glides through the cosmic expanse. Its engines emit a blue glow, symbolizing the advanced propulsion technology that powers its journey. Jupiter, the largest planet in our solar system, looms in the distance, a magnificent giant with its swirling bands of clouds and iconic Great Red Spot. The spaceship's trajectory takes it closer to the gas giant, offering passengers and crew members a spectacular view of the planet's immense beauty.

Depict a futuristic spaceship heading towards the Moon. The spacecraft is modern and sophisticated, with a streamlined body and glowing propulsion systems that propel it gracefully through the cosmic void. The spaceship's exterior is a testament to advanced aerospace engineering, featuring a polished metallic surface that gleams in the soft moonlight as it approaches its lunar destination. Its streamlined design and state-of-the-art propulsion systems ensure a precise and controlled journey.

Design an image of a futuristic spaceship traveling towards Earth. The spaceship is a stunning example of advanced technology, featuring a sleek, metallic exterior with glowing propulsion engines that emit a brilliant blue light, symbolizing its powerful thrust. The spaceship's streamlined design allows it to glide gracefully through the cosmic expanse. Its polished metallic surfaces reflect the Earth below, creating a striking contrast against the backdrop of space. The spaceship's advanced propulsion systems ensure a precise and controlled journey as it approaches the planet.

Create an image of a high-tech spaceship embarking on a voyage towards the Sun. The spacecraft is a masterpiece of cutting-edge design, featuring a streamlined and sleek exterior that reflects the brilliance of the Sun and the stars in the cosmic backdrop. The spaceship's advanced propulsion systems emit a vibrant, fiery glow as they propel it toward the Sun. Its polished metallic surfaces gleam with the promise of a mission that will unlock the mysteries of our closest star.

Visualize a daring scene where a futuristic spaceship is diving into a black hole. The spaceship is designed with a sleek, aerodynamic form, illuminated by a mesmerizing blue glow that emanates from its advanced propulsion systems. The spaceship's exterior is a testament to cutting-edge aerospace engineering, featuring a polished metallic surface that gleams in the cosmic darkness. Its streamlined design signifies its capability to navigate the extreme gravitational forces surrounding the black hole.

The bridge of a spacecraft is a futuristic marvel, featuring a panoramic view of the cosmic expanse beyond. At the center of the bridge, two skilled pilots are at the helm—a South Asian woman and a Black man. They work in perfect synchronization to navigate the spacecraft through the vastness of space. In front of them, a state-of-the-art console displays a mesmerizing array of holographic interfaces and controls. These interfaces llow the pilots to interact seamlessly with the spacecraft's advanced systems, from navigation to communication and monitoring.

The bridge of the spacecraft is indeed a futuristic marvel, a testament to the advancements of space technology. The panoramic view through the expansive windows provides an awe-inspiring vista of the cosmic expanse, with stars, planets, and distant galaxies stretching out into infinity. At the heart of the bridge, two highly skilled pilots take their positions—a South Asian woman and a Black man, representing diversity and excellence in space exploration. They sit side by side, working in perfect synchronization to guide the spacecraft through the vastness of space.

La nostra vita con i robot

Here is a visual representation of how artificial intelligence might conceptualize humanity: In a conceptual representation, you might envision AI viewing humanity as data points, interconnected networks, and patterns of behavior. It would see humans as nodes in a vast social network, with lines connecting them to represent relationships and interactions. Data streams and information flows between these nodes, symbolizing the exchange of knowledge and communication among individuals.

The image captures a serene moment with two robots, one with a masculine build and the other with feminine features, sitting on a lush grassy field. The setting is idyllic, with a clear blue sky overhead and gentle rolling hills in the distance. The robots are positioned close to each other, their metallic bodies reflecting the warm sunlight. The masculine robot has a strong and sturdy design, with angular features and a powerful presence. It sits with a relaxed posture, its mechanical limbs at ease, and its face displaying a sense of contentment. The feminine robot, in contrast, has a more graceful and elegant appearance, with curved lines and delicate details. It also sits in a relaxed position, and its facial features convey a sense of tranquility.

A captivating scene unfolds in a vast desert expanse where two robots, one with masculine features and the other with feminine, stand amidst the towering sand dunes. The setting is otherworldly, with the golden dunes stretching as far as the eye can see and the intense heat of the sun casting long shadows. The masculine robot has a robust and imposing design, with angular lines and a sturdy build. It stands confidently in the shifting sands, its mechanical limbs firmly planted in the desert floor. The feminine robot, on the other hand, possesses a graceful and elegant appearance, with curved contours and intricate details. It stands beside the masculine robot, mirroring its poise.

A serene image unfolds on a picturesque country lane, where two couples stand face to face. The human couple consists of a man and a woman, dressed in casual attire, with smiles on their faces as they gaze at each other. They radiate warmth and affection. The robotic couple, on the other hand, is a testament to advanced technology and design. The male robot has a strong and masculine build, with sleek, metallic features and a serene expression. The female robot exudes grace and elegance, with delicate details and a composed demeanor. Both robots stand with perfect poise, mirroring the human couple's posture.

A charming image unfolds as two couples, one human and one robotic, walk side by side on a scenic country road. The human couple consists of a woman of Southeast Asian descent and a man of African descent, both dressed casually and holding hands with smiles on their faces. The robotic couple is a marvel of modern design and technology. The male robot has a strong and masculine build, with sleek, metallic features that complement his confident stride. The female robot exudes grace and elegance, with delicate details and a serene expression. They walk in perfect harmony with the human couple, symbolizing a seamless integration of technology and humanity.

A lively image comes to life on a bustling city street, where two couples walk together, one human and one robotic. The human couple is a blend of cultures, with a man of South Asian descent and a woman of European descent. They walk hand in hand, smiling and chatting, as they navigate the vibrant urban environment. The robotic couple is a testament to advanced technology and design. The male robot stands tall and sturdy, with a sleek metallic exterior and a confident stride. The female robot exudes elegance and sophistication, with graceful lines and intricate details. They walk alongside the human couple, seamlessly blending into the cityscape.

A futuristic urban scene unfolds with diverse people bustling about, each wearing advanced technology clothing and gadgets. In the center of this vibrant cityscape, an Asian woman stands out, wearing a sleek, silver bodysuit that seamlessly integrates with her body. The woman's futuristic attire is adorned with intricate patterns of luminescent lines that flow across the fabric, giving her an ethereal appearance. The clothing is not just for aesthetics but serves practical purposes, enhancing her physical capabilities and providing real-time information through a heads-up display.

Cartoni animati

A heartwarming stylized cartoon image depicts a diverse group of children playing together with a toy train. The scene is filled with joy and unity, showcasing a Black girl, an Asian boy, a Hispanic boy, and a White boy, all happily engaged in their play. The children are gathered around the colorful toy train set on a vibrant, grassy meadow. They work together to assemble the tracks, connecting pieces with enthusiasm and teamwork. Laughter fills the air as they take turns pushing the toy train along the winding tracks, their faces lit up with smiles.

A stylized cartoon image of a diverse group of children playing with stuffed toys. The scene includes a Black girl, an Asian boy, a Hispanic boy, and a White girl, each engaged in a joyful playtime activity. They are sitting in a circle on a colorful rug, with the sun streaming in through a nearby window, casting a warm glow on their cheerful faces. The room is filled with laughter and the soft sounds of their imaginative adventures, as they each bring their favorite stuffed animals to life, sharing stories and creating a world of their own in a display of friendship and harmony.

In this heartwarming and inclusive stylized cartoon image, a larger group of children gathers to play in a vibrant and welcoming room. The scene celebrates diversity and unity as children from various backgrounds come together for a joyful playtime. The room itself is a haven of fun and creativity, filled with colorful toys, books, and playful decorations that create a warm and inviting atmosphere. Children of different ages and ethnicities engage in a variety of activities that showcase their unique interests and talents. Some build intricate structures with blocks, others immerse themselves in the world of books, a few express their creativity through drawing, and some share laughter while cuddling with stuffed animals.

A stylized cartoon image showing a diverse group of children playing on a Persian rug. The children include a Middle Eastern boy and a South Asian girl, along with their friends of various descents. They are surrounded by colorful toys and games, their faces alight with joy and laughter. The intricate designs of the rug add a vibrant backdrop to their imaginative play, with patterns that seem to dance under the playful chaos of toys. The air is filled with the sounds of their cheerful banter and the quiet hum of contentment as they build castles, share stories, and enjoy a moment of pure, uninhibited childhood.

In this vibrant and heartwarming cartoon image, a group of children from diverse backgrounds comes together at a lively park to enjoy a day of play with dogs. The scene is filled with joy and unity as children of various descents share in the delights of outdoor fun.

The park is a lively and inviting place, featuring lush green grass, towering trees, and a clear blue sky above. Children of different ethnicities and descents interact with dogs of various breeds, emphasizing the universal love for animals. A Middle Eastern girl gleefully tosses a frisbee to a playful dog, while a Black boy joins in the excitement by running alongside a Labrador retriever. An Asian boy engages in a game of fetch with a Border Collie, and a Hispanic girl shares a tender moment with her fluffy Pomeranian.

In this vibrant and heartwarming cartoon image, a group of children from diverse backgrounds gathers at a lively park to play with dogs. The scene is filled with joy and unity as children of various descents come together for a delightful day. The park is a colorful and inviting place, with green grass, tall trees, and a bright blue sky overhead. Children of different ethnicities and descents interact with dogs of various breeds, showcasing the universal love for animals. A Middle Eastern girl tosses a frisbee, and a Black boy runs alongside a playful Labrador retriever. An Asian boy enjoys a game of fetch with a Border Collie, while a Hispanic girl shares a moment of affection with her fluffy Pomeranian.

In this captivating cartoon image, a diverse group of children is depicted playing with dogs, seamlessly superimposed on a photorealistic background of a breathtaking natural setting. The scene combines the charm of cartoon characters with the beauty of a real-world environment. The natural background features a stunning landscape with lush greenery, a serene river, majestic mountains in the distance, and a clear blue sky overhead. The attention to detail in the photorealistic backdrop creates a sense of immersion in the scene.

In this photorealistic image, children from diverse ethnic backgrounds are captured playing with real dogs in a stunning natural setting. The scene beautifully portrays the universal joy of childhood and the love for animals. The natural backdrop is breathtaking, featuring a lush forest with towering trees, a tranquil river flowing gently, and a clear blue sky overhead. The attention to detail in the photorealistic environment creates a sense of immersion in the scene.

In this photorealistic image, a diverse group of children is captured playing with a single cat in a vibrant and cheerful children's bedroom. The room is filled with bright colors and playful decorations, creating a welcoming and joyful atmosphere. The walls of the bedroom are adorned with colorful posters and artwork, reflecting the children's personalities and interests. Stuffed animals and toys are scattered around the room, adding to the sense of playfulness.

In this photorealistic image, children of diverse descents are depicted playing with cats inside a cozy apartment. The scene captures the warmth and joy of their interactions with these feline companions. The apartment setting is inviting, with soft lighting, comfortable furniture, and a homey atmosphere. Children of various ethnicities and descents engage in different activities with the cats. A South Asian girl sits on the floor, gently brushing a cat's fur with a smile of affection. Nearby, a Black boy plays with a playful kitten using a feather toy, while an Asian boy watches with delight as his cat playfully paws at a ball of yarn. A Hispanic girl sits on a chair, cradling a contented cat in her lap.

In this photorealistic image, children from diverse backgrounds are depicted playing with a single cat in a cozy apartment. The scene exudes warmth and joy, showcasing the universal love for pets and the bonds formed through playful interactions. The apartment setting is inviting, with soft lighting, comfortable furniture, and a homely atmosphere. Children of different ethnicities and descents gather together to enjoy their time with the cat. A South Asian girl, with a warm smile, gently pets the cat's fur, while a Middle Eastern boy engages in playful antics with a feather toy, making the cat leap with excitement.

Appendice con immagini varie

In this scene, a man and a woman are engaged in a heated argument within a kitchen setting. The tension in the air is palpable, and the surroundings reflect the intensity of their disagreement.
The kitchen is well-equipped with modern kitchenware and furniture. Countertops, cabinets, and appliances provide the backdrop for this emotional confrontation. The cluttered countertops and scattered kitchen utensils indicate a disruption in the normally orderly space.

In this heartwarming scene, an elderly man and woman are depicted adoring each other in the comfortable and cozy setting of their living room. The atmosphere radiates warmth and love, capturing a beautiful moment in their enduring relationship. The living room is tastefully decorated with comfortable furniture, soft lighting, and personal mementos that reflect the couple's shared history. A warm, inviting fireplace may be gently glowing in the background, adding to the coziness of the space.

In this scene, a father is scolding his young son in a small, child-friendly bedroom. The walls of the room are painted in bright, cheerful colors, and the furniture is kid-sized and playful, creating a typical children's bedroom atmosphere. The center of attention in the room is the broken toy, which lies on the floor, partially disassembled. The father stands near the toy, while the young son, with a guilty and remorseful expression, sits on the floor beside it.

In this image, a figure resembling a historical theoretical physicist from the early 20th century is depicted deep in thought in front of a blackboard filled with complex equations and mathematical notations. The atmosphere is one of intellectual intensity and scientific exploration. The physicist, with a focused expression and wearing attire reminiscent of the era, stands or sits in front of the blackboard. Their posture exudes a sense of contemplation and concentration as they grapple with the intricate equations that symbolize their groundbreaking work in the field of theoretical physics.

In this historical scene, a figure resembling a 17th-century physicist is depicted sitting under an apple tree. The setting harkens back to a pivotal moment in scientific history, often associated with Sir Isaac Newton. The physicist, dressed in attire typical of the 17th century, is seated on the ground beneath the apple tree. He is deep in thought, contemplating the natural world around him. Meanwhile, a young boy stands beside him, looking up at the apple tree with a sense of curiosity and wonder.

In this dynamic scene, a large aircraft carrier is depicted floating on the expansive ocean, with a clear and pristine sky stretching above. The aircraft carrier's massive deck dominates the foreground, and it serves as a platform for various aircraft operations. An airplane is captured in the midst of landing on the carrier deck, showcasing the precision and skill required in such a high-stakes maneuver. The landing gear of the aircraft is in the process of making contact with the deck, demonstrating the crucial moment when the aircraft must achieve a controlled landing.

In this imaginative scene, a spacecraft is shown in the process of landing on the historic Giza Plateau in Egypt. The iconic backdrop features the Great Pyramids and the Sphinx, two of the world's most famous ancient landmarks. The spacecraft, with its sleek and futuristic design, descends gently towards the plateau, its landing gear engaged. The scene is bathed in the soft glow of the setting sun, creating a warm and otherworldly atmosphere. The advanced technology of the spacecraft contrasts with the timeless grandeur of the pyramids and the enigmatic Sphinx.

In this dynamic and awe-inspiring scene, the massive spaceship from the previous image is depicted making a majestic landing on a desolate alien desert planet. The planet's landscape is barren, with vast expanses of sand and rocky terrain stretching as far as the eye can see. The spaceship's landing gears are fully deployed, and it gently descends toward the planet's surface, kicking up a swirl of dust and sand as it approaches

In this thought-provoking image, a scientist's head is depicted in a moment of deep contemplation. The scientist's features are marked by a life of thought and dedication to their work. The scientist, likely in a laboratory or study, is immersed in their thoughts, with furrowed brows and a focused expression. Their eyes may be fixed on a complex equation or a specimen under a microscope, reflecting the intensity of their intellectual pursuits.

www.ingramcontent.com/pod-product-compliance
Lightning Source LLC
LaVergne TN
LVHW031711230826
846093LV00022B/505